Praise for Jad Josey's
Cast Your Longest Shadow

With gut-wrenching perception, Jad Josey's *Cast Your Longest Shadow* meets the ideal purpose of a chapbook: to focus the consciousness of a speaker bearing with raw honesty upon a unified subject. For a remarkably moving experience, I highly recommend that readers absorb the whole book in one sitting. In an exceptional sequence of stand-alone poems, Josey renders the existential displacement of an individual whose life has been rent by the sudden reality of lost love. Classic, universal, sharply contemporary, and formally surprising, these whip-smart, keenly imagined poems rise out of a breached mire to produce a blues verse that enthralls.

—Kevin Clark, author of *The Consecrations* and *Self-Portrait with Expletives*

"I've learned not to be haunted... in the between spaces" Jad Josey writes in *Cast Your Longest Shadow*, a book which flings itself simultaneously into lived-in images of nature and the narrator's ever-expanding interiority. Here, every poem is a "between space" built from liminality, love, and loam, a place where feeling brushes up against the stars only to then plummet into the ocean to glide alongside an ancient shark. While it is impossible to squeeze everything into a poem, in these perfectly-crafted lines, Josey has certainly come close. "When you have / nothing left, you still have / everything you carry."

—Todd Dillard, author of *Ways We Vanish*

CAST YOUR LONGEST SHADOW

Chestnut Review Chapbooks, an imprint of Chestnut Review LLC
Ithaca, New York

https://chestnutreview.com
ISBN: 978-1-965158-18-0

CAST YOUR LONGEST SHADOW

JAD JOSEY

Chestnut Review Chapbooks

CONTENTS

Inflorescence

What if the first person to sit
beneath the fig tree hadn't plucked

that strange flower, hadn't dared to die—
had instead carried on without

tasting its sweetness. We stare through
windows at horses in the pasture,

imagine saddle and bit and reins
instead of freedom. The grass

bows in the breeze, pregnant heads
nodding, and we were always

all of this without ever knowing:
flowers growing inward,

never noticing the light until
the teeth come tearing through.

Cadaver

I don't want anyone to carry my casket.
I would rather be nailed to a tree. Summon
a thunderstorm or an unlikely
summer hail, even the late season
fog. Just not the wind, please, not
the goddamn wind carrying
those egrets higher than they meant to go,
secrets and regrets all aloft.
I want to come apart, to undo
and be undone, to slide
neither reckless nor resolved into
a thousand hungry mouths, be
claimed by whatever is willing
to fight for the very last thing I can give.

Enormous as Want

Picture a sky enormous as want,
a road stretched out in the shape of your longing.

Scrub brush and ice plant, thin trails cut
by deer. There are stars behind this wilderness

we will never see. There are stars behind
those stars we cannot imagine. When something comes

you must let it. The light has traveled so far
to find you here. Meet it where you can.

Cast your longest shadow and reach
your umbral hand toward mine.

We can walk the road together beneath
the insatiable sky, shadow fingers touching.

Our fingers touching, too.

Different than Sad

I bought a mill for my stand mixer
to grind my own wheat berries. The wind
from the south has tipped the fence. I never
wanted to see into the neighbor's kitchen.
The man is drinking red wine and talking
with his hands. Her mouth has barely
moved. She looks different than sad,
the tired that comes from asking
unanswered. I have cucumbers fermenting
in the cupboard. The jar bubbles and it
feels like progress. My hands have done
something. The wheat berries remain
unmilled—but not for much longer,
I tell myself.

Kind of World

There are no sunk costs in love,
though you might be a boat, though
you may be

sinking, treasure chest destined
to find the bottom
of the sea,

of the blue-black lake,
the mattress where the scoop
of your body stays well past

your leaving. The cost
might be high. The price may be
unforgivable.

Your heart. A golden locket.
The friend you left wondering.
All the time you spent.

When you have
nothing left, you still have
everything you carry.

What should sink suddenly rises,
surprising even you.
What kind of world is this

where a heart can just stop beating?

Everything Goes

Learning again to unwind time
from its branches. The lupine in the hills

dead-headed and dog-tired, barely a spark
would send us up in flames. I have grown

another day wiser, but of course
this isn't true. I have grown another day

closer to coldness, have taken up the fight
to keep tenderness near the surface

while the trees shrug off the summer fog
and the Matilija poppies brown

like eggs too long in the pan. Even the bees
have forgotten the ecstasy

of mere days ago.

Still Life of a Desk

No framed photos, no coffee mugs stained with time,
but a small chunk of sandstone adorned with large googly eyes,

the whites yellowed beneath the sun that streams in, right eye
cracked from a fall—from a life lived, you might have said.

You might have said that, but you are not here.
I have waited so long for the one who returns, who

closes the door and then, in a rush of whatever compulsion
chambers the heart, reappears at the door to knock quietly,

but urgently, and—when I open it—looks at me with broken,
yellowed eyes that promise

I will never leave again, not from this life lived, not ever.

I've Learned to Pinch an Orange

from the tree with barely a leaf unstilled,
learned to stop saying to anyone

who'd listen that I would have
stayed with you until the end.

I've learned not to be haunted by the pith,
bitter white in the between spaces,

that absence of color I find in my waking
dreams, eyes open wide but the world

all dimmed. Breath of wind, stir those leaves while
I pace the dirt roads, cleave open

the blossoms before I arrive.

They All Do the Same

There is often a barn owl outside my bathroom window.
His eyes are usually closed when I'm in there,
when I've stripped off my clothing
to shower or to cry alone. He was there
before my pieces no longer fit, before the
together changed. He turns his head away.
They all do the same. I wonder where he goes
when it rains. He doesn't remember me
the old way. The rain falls and he vanishes.
And then one morning he doesn't.
The raindrops ring his feathers like
jewels. I am naked and his eyes are open.
We're back together in a new way.
I'm still learning. Of course, he
already knows.

The Birds Lit Out

and they lit out in perfect unison, the sound of it like vinegar
when the tongue divines sugar,

that treachery like the night I pedaled to your house,
all those miles on the dust-swirled county road,

rode beneath a waning moon so slight it barely shone, rode
to stand beneath your bedroom window

and whistle our secret melody,
rode to stand beneath that feeble crescent

to wait and
keep waiting

until you silhouetted the window.

You pulled the curtain closed so slowly
it could not have been a mistake, and

this moment—like the moment the birds
flush with no warning—a taste

on my tongue so different than what
my soft animal body expects,

this moment shudders like I shuddered then,
rain tapping softly as the waxing moon dips into darkness.

What Rises Is Something Different

The fluttering is not butterflies.
Not those iridescent wings
beating from my belly,
clambering into my chest.

I rarely trust the good things in this world.

I hold them like a single
piece of tissue paper in
the wind: fingers fastened
to the corner, certain the squall
will get its work done.

When I ask

Are you okay, love? Is
everything between us
real, love? What would the
breaking sound like, love?

those are not butterflies
rising from my throat
but fireflies whose light
has been smothered,
shadows crashing up
and up and finally out

somehow still alive
but not for long, probably.

And Then It Was June

I lost a whole season to
the tremor inside my heart. Some of it

was wind, no great loss, but some
was slant light strafing

stacked clouds, columns collected atop
a cobalt sea—what joy I would have found there.

I was shaken like a snow globe, sudden
storm and viscous uncertainty, staggering

through that smooth wilderness in search of
stillness. I emerged to a chorus of frogs

erupting into silence at my unsteady step,
blue whale spouts silhouetted against a summer curtain of

fog. What longing happened while I was gone?
Did you spot a comet streaking the sky?

Come, sit with me. Let me close my eyes
while you tell me who you are now.

Independence Day, Coffee Cold

The mosquitoes are painting
silhouettes of themselves
in the late afternoon sun,
and it is beautiful beside the
bay, everything is gorgeous
before the sun swallows itself,
this small, unsteady pier
macabre with the bloodletting
while the blasting of bottle rockets
and black cats bounces across
the water, cormorants scattering
skyward like we wish we could,
all of it suddenly sullen and
morose. The coffee has
gone cold. It splashes quietly
into the dark water. We were
always these animals, shivering
after the sun, seeking warmth
that cannot come from
fire or freedom.

Aftermath

The scorpion emerged from shadow, big enough to matter
and small enough to matter, the world

a dram conspiring between floated spirits,
the grain alcohol, the smoky absinthe,

the way my mother is eleven-hundred miles close
and yours is eight miles and nearly vanished,

an apparition you didn't intend to summon, though
you might have wished her gone one sticky-hot September evening,

never divining the prophet you'd become, never mind how small her
hand felt in your palm, her heart barely here, the goodbye gone.

The scorpion scuttled onto my foot, and I waited. Waited
for the pricking poison, waited for what comes before the aftermath.

Waited for something small to bring me to my knees again.

The Moment I Knew

I was making coffee
for her
but I was thinking
about you.

Not Bruise, Not Eggplant

I wandered into the garden
to find what there was to reap and—
like a bird waking from a dream
of silver-edged clouds pocking sea-blue skies
to find its feathers vanished—
I found only dark earth, earth dark with dew.
Later you are making tea,
water rumbling over flame,
and the gloaming is too loud, too quiet,
you cannot be sure. The horizon is a purple
you've never seen before—not a bruise,
not a goddamn eggplant. Pull me into you.
There is hardly any time left in this world.

Kneading Dough on a Moonless Night

as one does when the stars are falling from the sky.

There was no rain. The countertop was dusty with flour, rosemary,
psilocybin powder. Was no rain. Were no rain. We need to talk about

the singularity of raindrops. You write couplets better than anyone
I know. I bake bread with dark brown ears and when have you

ever loved anything more. No reason to dream about
what you might love in a later season. Sometimes a dream undoes itself

with no sound. I have felled ghosts taller than these trees, I have
walked the dirt road under lightning light. You are new to this—not new

to this life, but new to *this* life, the one you've worried with a thumbnail,
peeled back to reveal. What I wouldn't give for some water, but

the mushrooms are drumming my blood and I won't be
swallowing for hours and the dough is elastic, it is

everything all the time now. We haven't been the same in weeks.
The metal hook is clean. The dough will rise and we will smolder

and I will sing you the song I haven't sung for anyone else.

Carousel

This spring wind is
a carousel: though
it does not turn, it
still delivers me
nowhere. I would
lubricate its grating
into silence. Even
the strong-winged
herons have given
up, not stoic in the
weary grass—no,
defeated, unflying,
tethered to the earth
like I am. We are
buzzing with the
static of an element
we would rather unsee,
the bending trees
our allies until we can
step from this ride
into a day with
nothing but the sun
at our backs.

The Oldest Living Thing in the World

Sea birds will skim their wings
across the broken mirror
of the sea, and you will not be there

to see it. A shark glides through the gelid ocean
while three centuries turn. She is the only time
that matters. Wing and dorsal tips catch

the same falling light. You are made of vagaries
and there are stars down to the horizon line.
There are stars beneath your feet even now.

When the light before you
becomes the light behind you,
nothing will change.

Everything will be different.

Both of Us Standing

at the edge of the sea,
the edge and something more.

Wild, these wildflowers
turned away from the wind. Your fingers

too cold to find mine. There was always
something missing too quiet to say.

Let's turn the dark matter into the light,
unbury it from our hearts where we can

use it later. After. The pelicans are pushing north,
wings wide above the wind that moves us.

We hide our faces while they ride.

Plow and Burrow

These bodies. Sometimes they refuse
to work and still we are beautiful. Last week
you pulled a carrot from the earth, a carrot
with two bodies. Two heads, you said, but I knew
you meant bodies because of the way
you sometimes want to plow your head and
burrow deep into your body.

You used to live next to a windmill.
How you wished it would turn.
You said it almost every day, hands
sweeping, palms turning. You the thing,
but the thing something different.
Like the way you can remember
the shape of a shadow though the
sun has set a hundred times since.

You flip the light switch
on and off a hundred times. Work,
you say, work and work and work.
Somewhere a power line is arcing.
A loose curl falls across your face. Someone
is pulling on leather gloves
so the bulb can show you to me.

Letter to the Pacific

The humpbacks have taken to the air—
bodies untethered from gravity in an unexpected
zenith—while messengers of light climb

down from the clouds. The whales are traveling
in the direction of my heart, guided by memory,
by sound, by invisible forces. I will follow them

across the sea, compass calibrated to a new
north. I am rising toward my own apogee. I can read
the map with eyes open or closed, trace the

contour lines of your body with fingers or tongue
or mind, traverse the ocean in darkness knowing
your lighthouse lamp is burning. You turn laughter

into longing, transform time into want, your
quiet alchemy collapsing distance while the whales
sing their secret songs beneath the blue.

We are watching the way joy meets the sky, the way
freedom finds its home. Every part of me
is on its way to you.

Everything from Now On

You collect shells
while I collect poetry.

They are not so different,
the two. One hazy beer

with the sun shining
through. The sky

a sky blue and
all the world everything

from now on. A frame
on the spool of time.

I realize I've been trying
not to love you all along.

Acknowledgments

I am grateful to the following publications and their respective editors for giving these poems such loving homes.

"Aftermath," *Okay Donkey Magazine*, April 2023.

"Both of Us Standing" and "Barely a Leaf Unstilled" [included as "I've Learned to Pinch an Orange"], *Terrain.org Magazine*, April 2023.

"Inflorescence," "Kind of World," and "Cadaver," *Frigg Magazine*, Issue 63, Fall/Winter 2024/25.

"Kneading Dough on a Moonless Night," *Lilac Magazine*, Summer 2022.

"Not Bruise, Not Eggplant," *Longleaf Review*, August 2021.

"The Birds Lit Out," *Bayou Magazine*, Issue 79, 2025.

About the Author

Jad Josey's work has appeared in *CutBank, Glimmer Train, Ninth Letter, Passages North*, and elsewhere. He has been nominated for the Pushcart Prize, Best of the Net, and Best Small Fictions, and his story, "It Finally Happened," was selected for inclusion in the Best Microfiction 2021 anthology. Jad is currently working on a novella, myriad poems, and a collection of short stories. Read more at www.jadjosey.com or reach out on Twitter @jadjosey.

www.ingramcontent.com/pod-product-compliance
Lightning Source LLC
Chambersburg PA
CBHW030529130626
46549CB00007B/3158